Sublime Nature

Sublime Nature

PHOTOGRAPHS THAT AWE & INSPIRE

Cristina Mittermeier

NATIONAL GEOGRAPHIC

WASHINGTON, D.C.

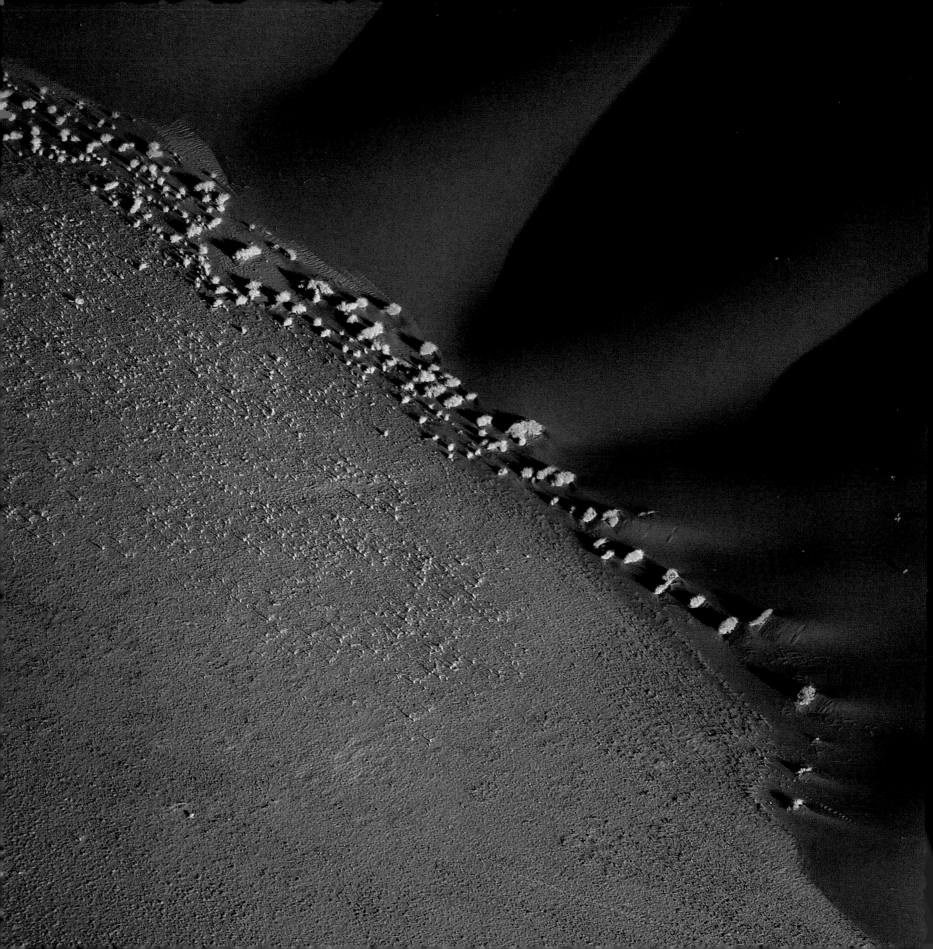

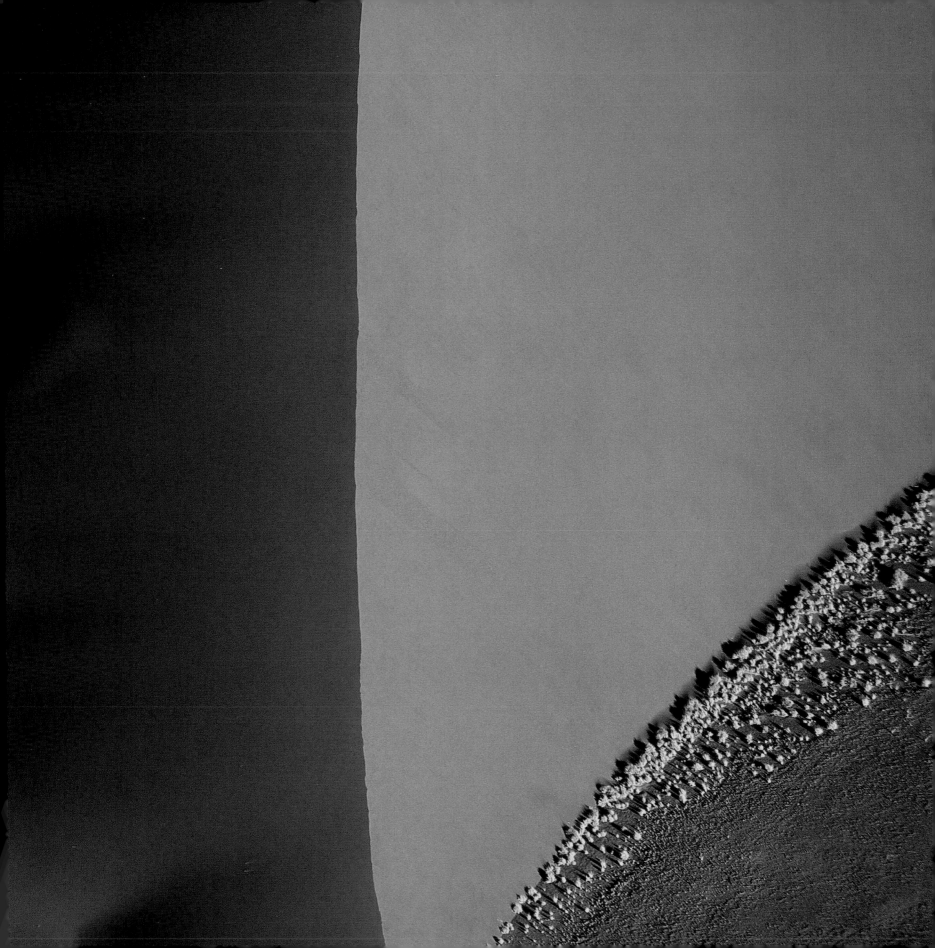

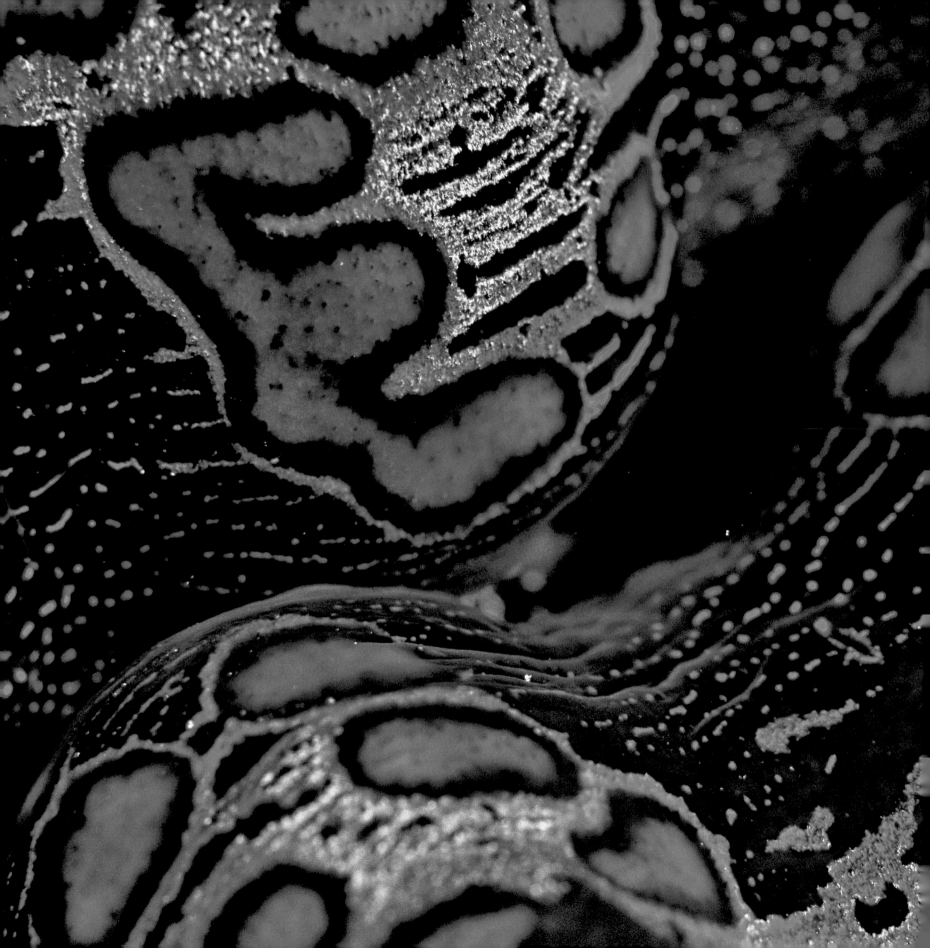

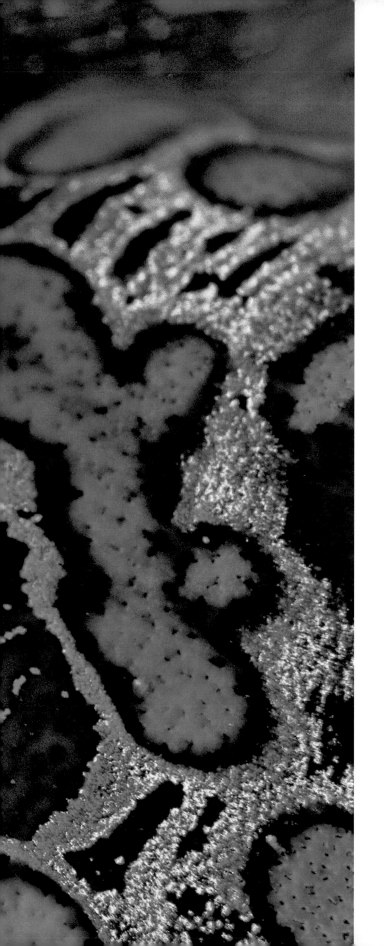

CONTENTS

Introduction 16

Awe 18

Grace 68

Joy 118

Peace 168

Afterword 218 | *Acknowledgments* 220

Illustrations Credits 221 | *Index of Photographers* 222

MARC HOFER *(2–3) Mist rises from a lake ringed with autumn colors in Seedort, Switzerland.* | FRANS LANTING *(4–5) Abstract forms emerge from this aerial view of the red-colored Sossusvlei dunes in the Namib-Naukluft National Park, Namibia.* | JOEL SARTORE *(6–7) Marching single file across the land, American bison roam the South Dakota plains at dusk.* | MARIA STENZEL *(8–9) Taking a break from their oceanic journey, chinstrap penguins rest atop a blue iceberg in the South Sandwich Islands, Antarctica.* | THEO ALLOFS *(10–11) A mesmerizing sunrise glows in the misty wetlands of the Pantanal, Brazil.* | TIM LAMAN *(12–13) Nature's magic can be found even in the smallest details, such as in this view of a giant clam's colorful mantle in Indonesia.* | CARR CLIFTON *(14–15) In British Columbia's pristine Skeena Mountains, the Nass River winds downstream from Nass Lake.*

Introduction

FOR MANY YEARS, PHOTOGRAPHY HAS BEEN MY OUTLET FOR ARTISTIC EXPRESSION AND the passport that has allowed me to immerse myself in the splendor and majesty of nature. I have traveled the planet with pen and camera, and through this journey I have come to understand the extraordinary power of picture-taking to serve as a universal language: unwritten and unspoken, yet fully capable of reaching all cultures with a hopeful message for all life on Earth. ⌒ And what a wonderful time to be an artist! With Earth's natural resources carelessly depleted or misused, we must find ways to remain optimistic. Over the course of my career I have witnessed photography's ability to shape perceptions, to help societies pause and reflect, and in some cases to inspire change. This book was born from my desire to share my love and respect for the natural world through the medium of photography and storytelling. ⌒ While choosing the images for this book, I had the opportunity to review thousands of photographs in National Geographic's collection, taken by some of the world's most talented photographers. Those pictures remind me of all the gifts nature provides us every day—water, food, shelter, medicine, fibers for clothing, fuels for warmth, and infinite means of recreation. Creativity, invention, religion, and cultural traditions also have sprung from nature and its intricate web. Almost every minute of the day involves our interaction with nature and its resources in small or large ways. ⌒ Last fall, when the leaves turned orange, I sat on the riverbank of the Fishing Branch River in northern Canada, patiently waiting for a special drama of nature to unfold. Grizzly bears, already fat with berries but not fat enough for their long winter slumber, were coming down from the hills. They were on a mission to feast on the big chum salmon that arrive annually in mighty pulses to their spawning grounds, their silvery, blue-green skin mirroring the color of the ocean, the starting place of their journey. ⌒ As the bears lumbered past only a few feet from where I was sitting, aware of me but indifferent to my presence, I was reminded that time spent in nature is a privilege, for our work lives are often intense and frenzied, focused on papers and computers. Witnessing nature's splendor through the view-finder of my camera has helped me to slow down and see things in a new, connected way. In order to see the river as a living thing, we must slow down enough to experience that presence in our own souls. When we are one with the river, we no longer see it as a mere commodity that can fill commercial water bottles, put fish on dinner tables, or bring glory for bearskin trophies. Instead, we are able to reflect on its miraculous value and what we can do to involve others in our appreciation. ⌒ Being a photographer has allowed me to understand—and share with others—that all things in nature are part of one vast ecosystem. Unlike people, the Earth's waterways, wildlife, and forests have not removed themselves from the fabric of the whole by claiming a separate existence. My hope is that the images and essays in this book will inspire a stronger connection with the nature that lies within and around us: infinitely worthy of our deepest respect and care.

Cristina Mittermeier
Nanoose Bay, British Columbia, 2013

The richness I achieve comes from Nature,
the source of my inspiration.

—CLAUDE MONET

Nature fills our hearts with awe as it showcases its might in both subtle and spectacular ways. A primordial engine, always stunning, always humbling, nature transforms the land, shapes the course of rivers, births entire mountains, and creates new life. We stand captivated by the steely gray clouds of an approaching storm that rumble with menacing thunder. Lightning splitting the night with jagged white electricity terrifies and thrills us, just as the Grand Canyon's miraculous gorge and geology leave us speechless. In late summer, we marvel at the sheer beauty of the sea's last tide, when an opalescent moon brings in waves from distant places to crash against a rocky shoreline, while bathing the whole world in a shimmering silver stole. The bald eagle's mighty lift to dizzying heights pierces our hearts with awe. Poised as if suspended by an invisible thread, the tips of its feathers flutter as it turns and arcs through blue sky before plunging straight down to the river below—a sunlit glassy vein winding through the wilderness. How we wish to be that eagle for one moment of bliss—to feel its talons gripping the silvery fish, whose strong, scaly body glitters in the sun like jewel-encrusted armor. As part of this grand, mysterious, and remarkable world, we are surrounded with stories and sights that fill us with awe: the psychic powers of a shaman; the northern lights; nature's familial hold on the human spirit, reminding us that the thundering waterfall is much more than a commodity to exploit; and nature's eternal inspiration for mankind's artistic creativity and personal spirituality.

MAREE TOOGOOD
Polka dots and ornate tentacles decorate a colorful sea slug, one of the thousands of underwater denizens of the South Pacific.

FRANS LANTING
Painterly reflections break the mirrorlike surface of a waterhole as a herd of African elephants come to drink at dawn in Chobe National Park, Botswana.

preceding pages
MARC ADAMUS
Wispy clouds come to life as the last rays of light set Oregon's Great Basin desert ablaze.

There is a pleasure in the pathless woods,

There is a rapture on the lonely shore,

There is society, where none intrudes,

By the deep sea, and music in its roar:

I love not Man the less, but Nature more.

—LORD BYRON

I move
in the university
of the waves.

—PABLO NERUDA

PAUL NICKLEN
The frozen Arctic landscape is framed through the eye of a piece of glacier ice near Svalbard, Norway.

The most beautiful thing

we can experience is the mysterious.

It is the source of all true art and all science.

He to whom this emotion is a stranger,

who can no longer pause to wonder

and stand rapt in awe,

is as good as dead: his eyes are closed.

—ALBERT EINSTEIN

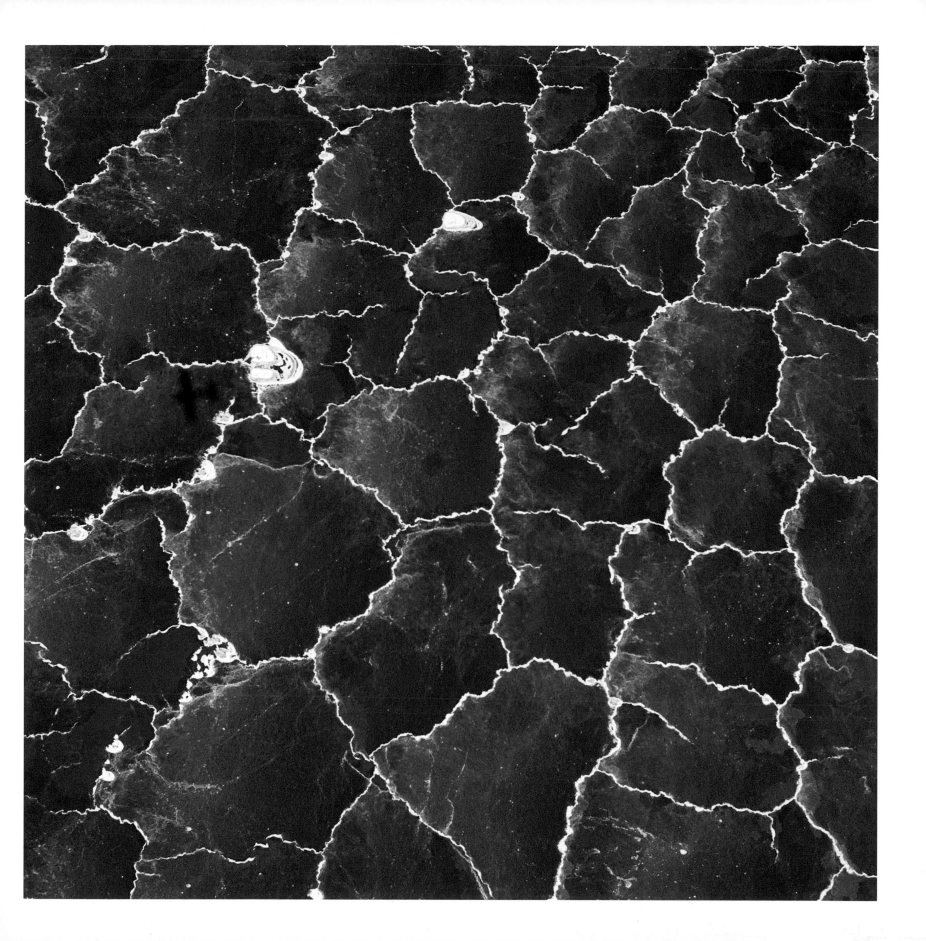

PAUL NICKLEN
Sea urchins with spectacular purple and orange spines populate God's Pocket Marine Provincial Park in British Columbia, Canada.

following pages
CARSTEN PETER
Massive selenite beams crisscross a cave in the Cave of Crystals in Chihuahua, Mexico.

MAURICIO HANDLER
Beautifully colored clouds pass over the rocky shore in Cape Elizabeth, Maine.

NICK FORNARO
The scales surrounding an iguana's eye resemble mosaics in a work of art.

What do we wish? —
To be whole.
To be complete.
Wildness reminds us what it means
to be human,
what we are connected to
rather than what we are separate from.

—TERRY TEMPEST WILLIAMS

ROBERT B. HAAS
After a small atoll island collapses, a magnificent
azure sinkhole remains, surrounded by coral,
in the Blue Hole Natural Monument, Lighthouse
Reef, Belize.

CARSTEN PETER
Awe-inspiring clouds, shaped like an alien ship,
swirl across the Texas Panhandle.

Simplicity

is the ultimate form

of sophistication.

—LEONARDO DA VINCI

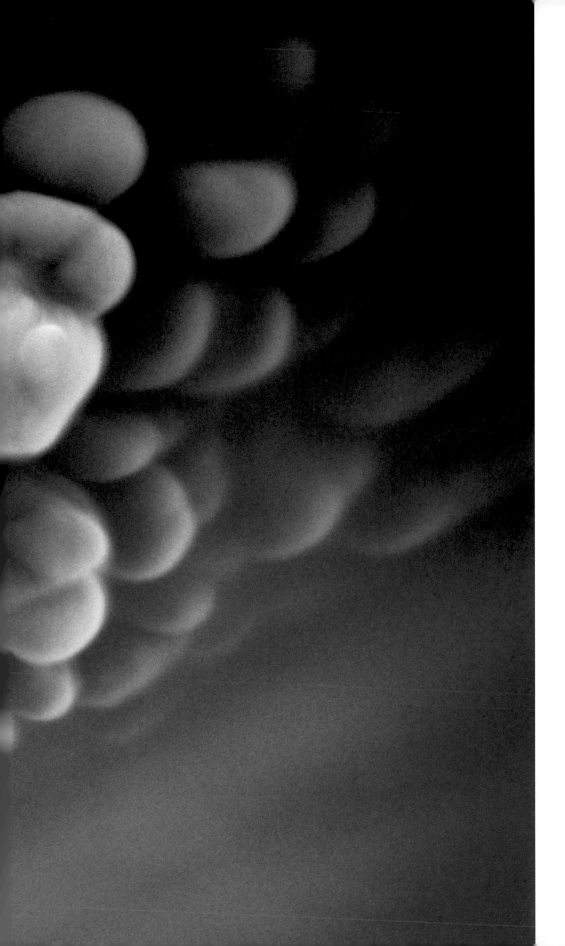

ALASKA STOCK
Cotton candy clouds dress up the sky as a volcano erupts in mighty steam bursts in the Kenai Peninsula, Alaska.

Slow and gentle, the Gibbon River reflects the sunset sky as it lazily meanders through a meadow in Yellowstone National Park, Wyoming.

Nature is a painting for us, day after day, pictures of infinite beauty.

—JOHN RUSKIN

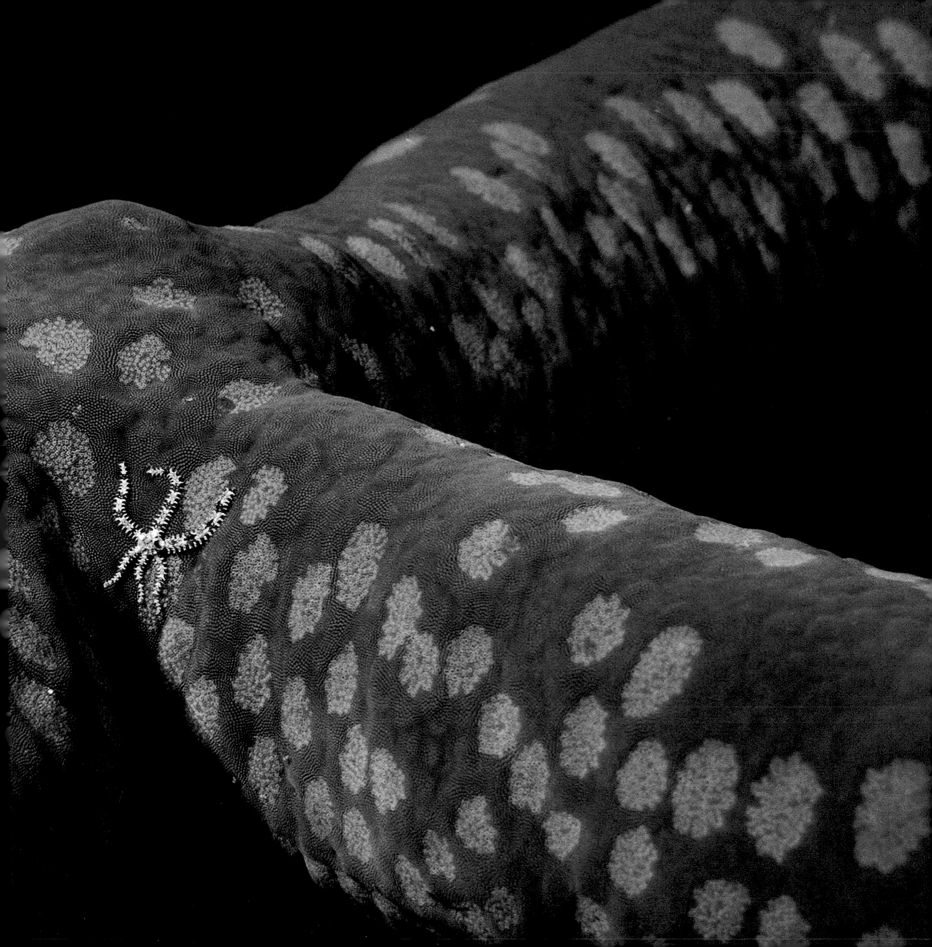

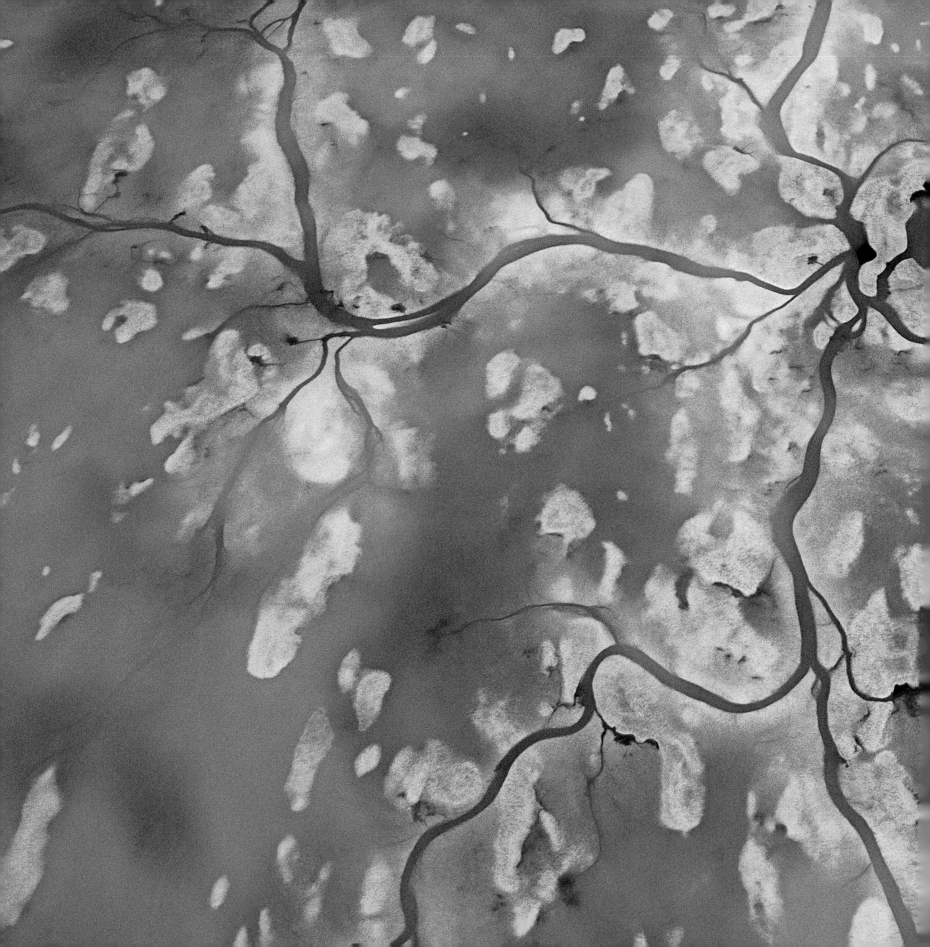

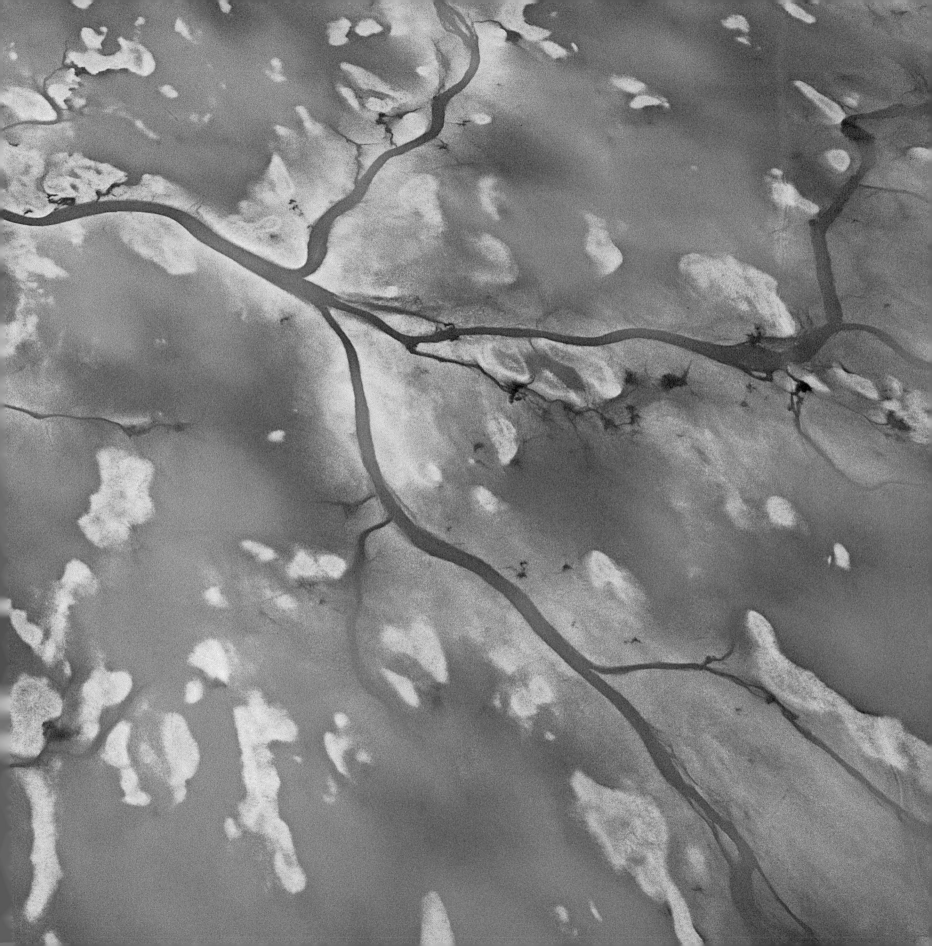

Each moment of the year
has its own beauty, ...
a picture which was
never before, and which
shall never be seen again.

—RALPH WALDO EMERSON

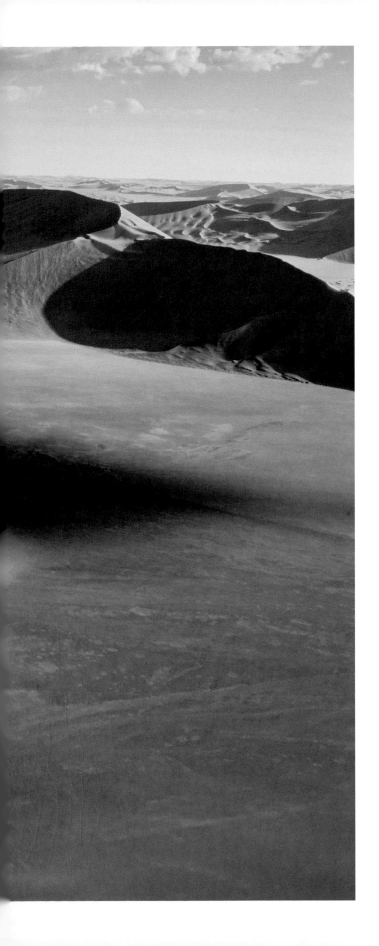

MICHAEL POLIZA
Rain revives the dry grasslands with vivid hues of green in the Sossusvlei dunes in Namib-Naukluft National Park, Namibia.

Serenity is not freedom
from the storm,
but peace amid the storm.

—ANONYMOUS

Those who contemplate the beauty of the earth
find reserves of strength that will endure
as long as life lasts.

—RACHEL CARSON

It is an often repeated adage: Look for grace in nature so that you can find it in yourself. The longing to understand our human place in the universe—to feel connected at a deep, primordial level with the natural world surrounding us—has from time immemorial filled our souls with grace. ⌒ Grace spreads its muted warmth when summer is about to end—when trees change their verdant coats for spectacular foliage of gold and fire. Like clockwork, following invisible cues from invisible maps, the southward-bound geese form a solid wedge in the sky that heralds autumn. Animals from squirrels to bears scour the ground and bushes for enough food to last the winter. ⌒ Grace remains within us during the last summer tide, when a sliver of moon draws the gently lapping sea to the shoreline's rocks. Cautious toes dip into the cool, inky sea as another day ends and the last coals of sunset glow on the water. The million specks of light that dance on the sparkling sea make it seem as if the northern lights had fallen into the water just so the ocean and our spirits could be joined together in grace. It is a tender seasonal parting with unspoken consensus to come together again when warm weather returns. ⌒ In autumn, a leaf-covered trail leads us into a forest pungent with fallen leaves, earthy humus, and scented pine. As we lift our eyes to the crowning web of treetops we experience the divine call of nature. We feel the sensation of a cathedral, a mosque, or any sanctuary, made all the more hallowed by the last leaves shimmering delicately, bravely in the sunset. John Donne's poetry comes to mind: "No spring nor summer beauty hath such grace, as I have seen in one autumnal face."

HANS STRAND
Ribbons of blue water wind down through the landscape of this river delta in Iceland.

ART WOLFE
Piercing the dusky gloom of a slot canyon, a ray of sunshine lights the smooth walls of this rocky chamber in Antelope Canyon, Arizona.

following pages
KOICHI KAMOSHIDA
Japanese cranes appear to dance in a snowy field in Kushiro, Hokkaido, Japan.

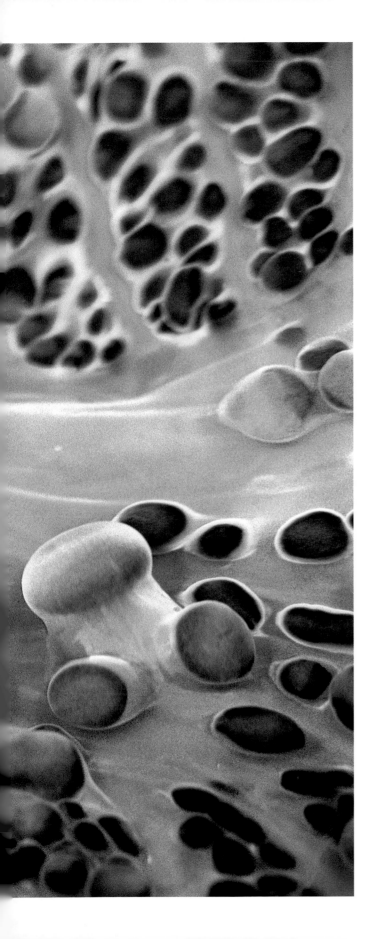

CHRIS NEWBERT
*A mushroom coral in soft green and blue hues
uses its unusual mouth to search for food particles
in its deep ocean home in the Solomon Islands.*

And the heart
of the great ocean
Sends a thrilling pulse
through me.

—HENRY WADSWORTH LONGFELLOW

IAN CAMERON
On a breezy summer day, a road winds
through a supernatural place of wonder:
the Quiraing, in the Isle of Skye, Scotland.

preceding pages
ART WOLFE
Wild purple lupines spread as far as the eye can
see in an Icelandic meadow.

MARC ADAMUS

Reminding us of nature's resiliency, two trees stand alone on the endless sands of the Empty Quarter desert in the United Arab Emirates.

Strive to have friends,
for life without friends
is like life on a desert island . . .
to find one real friend
in a lifetime is good fortune;
to keep him is a blessing.

—BALTASAR GRACIÁN

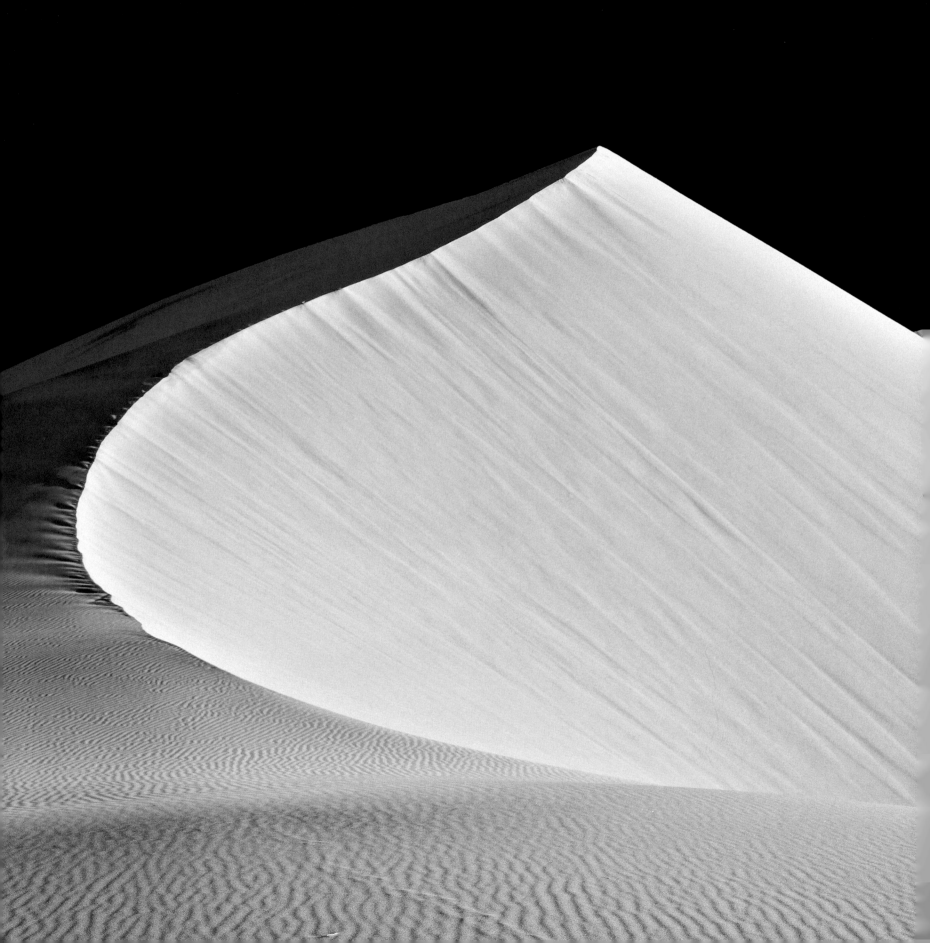

If you truly love Nature,

you will find

beauty everywhere.

—VINCENT VAN GOGH

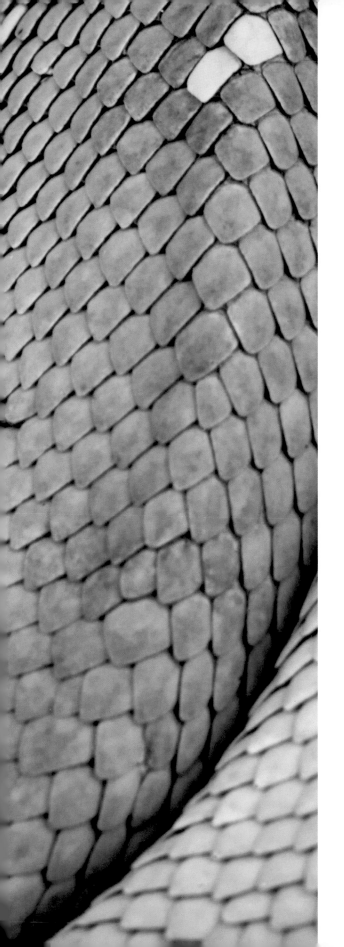

In Svalbard, Norway, the never ending ebb and flow of the ocean's currents dimples the face of glacier ice, creating astonishing designs.

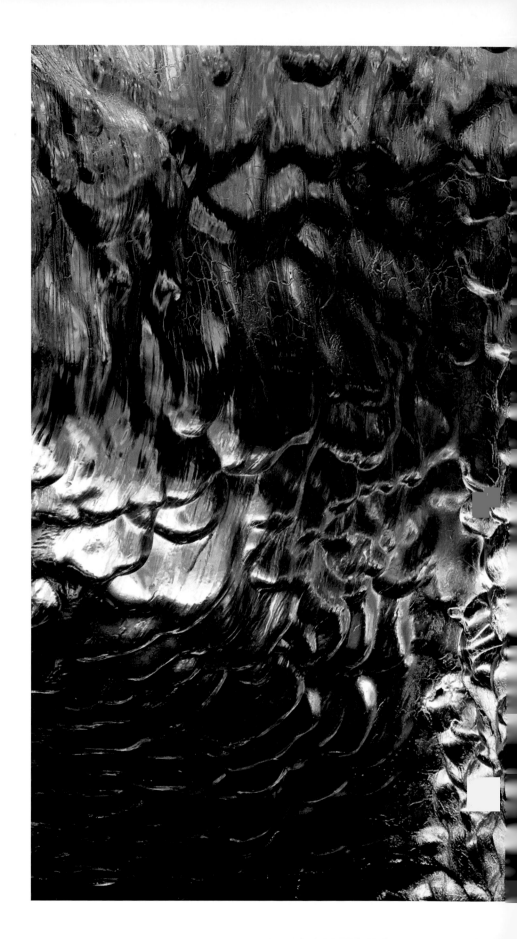

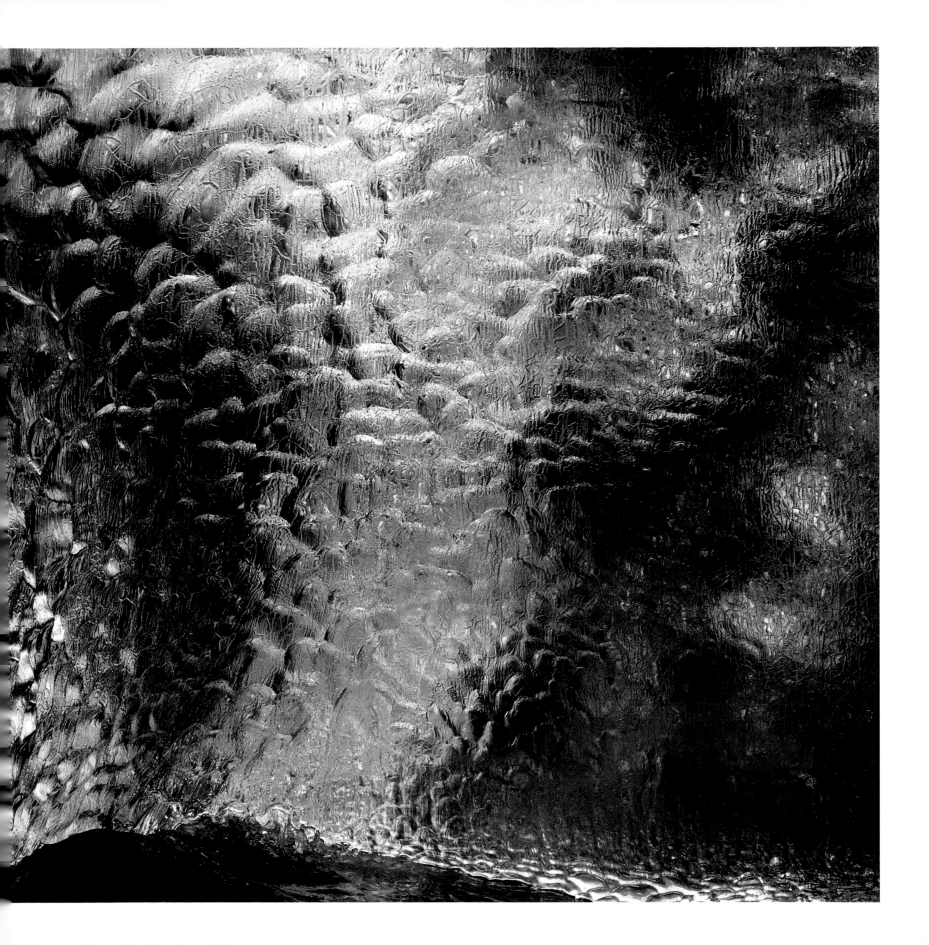

JOE PETERSBURGER
Just before its body breaks the water's surface, this kingfisher casts a perfect reflection of itself on mirrorlike water in Lábod, Hungary.

Reflect upon
your present blessings—
of which every man has many—
not your misfortunes,
of which all men have some.

—CHARLES DICKENS

LAURA CRAWFORD WILLIAMS
Taking flight in a mad flutter of wings and feathers, mallards create an artistic mosaic in Tuthill, South Dakota.

preceding pages
YVA MOMATIUK & JOHN EASTCOTT
The yellow glow of morning light creates a perfect mirror for Mount McKinley and the Alaska Range.

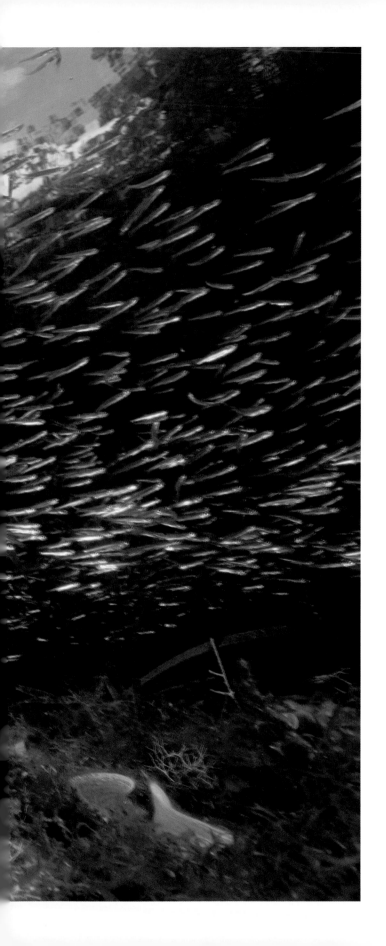

Have faith
in small things
because it is in them
that your strength lies.

—MOTHER TERESA

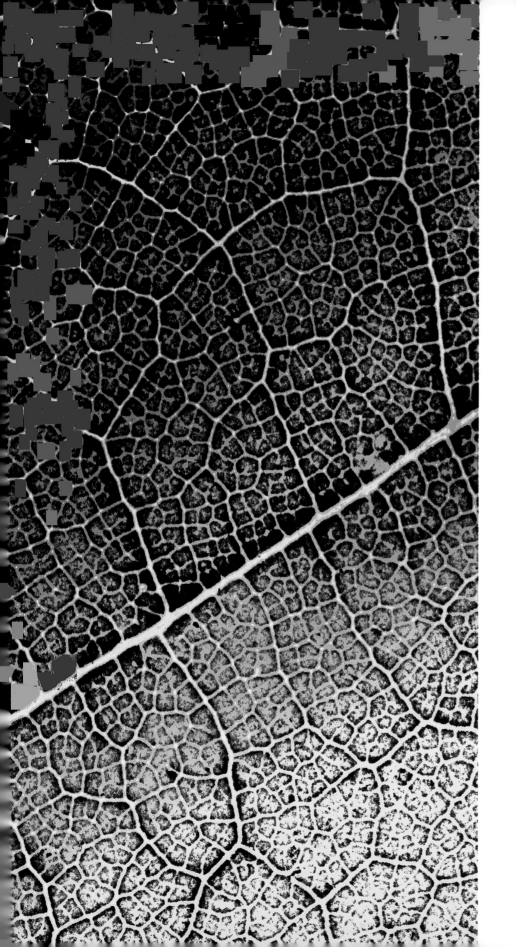

The beauty of nature lies in its endless variety of details, such as the veins of this fall maple leaf in the North Woods in Minnesota.

FRANS LANTING
A lovely, silent landscape is formed by dead camel thorn trees against a backdrop of the Sossusvlei dunes in Namib-Naukluft National Park, Namibia.

following pages
JOE PETERSBURGER
Flying at dizzying speed, a female bee-eater plucks a butterfly from the air to feed her chicks in Sáránd, Hungary.

The dawn of life
is like the dawn of day,
full of purity, visions,
and harmonies.

—FRANÇOIS-RENÉ DE CHATEAUBRIAND

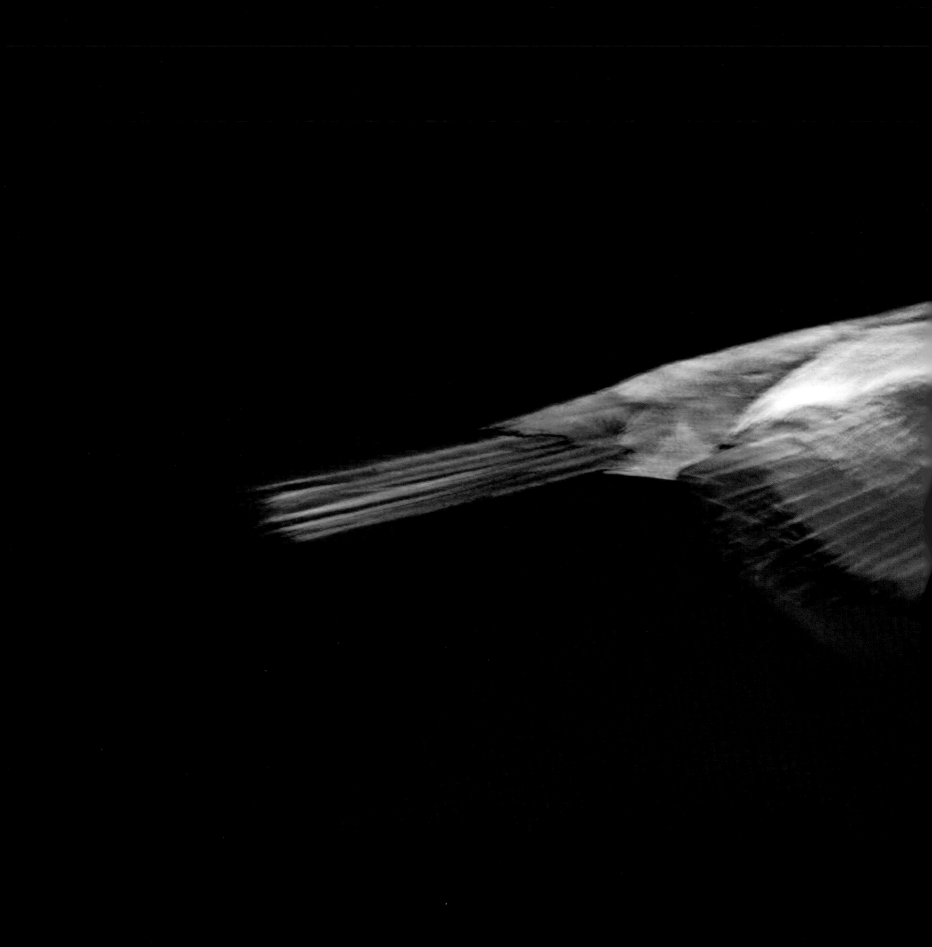

A grassy meadow dotted with golden tickseed
and foxtail barley sways in a breezy afternoon
near the Mormon Lake Basin in Arizona.

If you will stay close to nature,

to its simplicity,

to the small things hardly noticeable,

those things can unexpectedly become

great and immeasurable.

—RAINER MARIA RILKE

STEFAN AUSTERMÜHLE
In what resembles a graceful ballet move, four bottlenose dolphins vault in perfect synchronicity through the shallow waters at Chilca Beach, Peru.

TIM LAMAN
The twisted and contorted branches of an ancient bristlecone pine lie silently amid a snowy wooded landscape in California's White Mountains.

Nature does not hurry,
yet everything
is accomplished.

—LAO-TZU

All nature wears
one universal grin.

—HENRY FIELDING

reezes on a spring day, whether gusting over oceans or rustling through forests, have the power to lift our spirits with joy. The earth grounds our being, and the blue sky spurs our imagination to higher planes. To find happiness in nature is to discover that we are part of something larger than ourselves. Nature delivers joy to each of us in wonderful and unsuspecting ways. The serene beauty of rolling hills or the epic summit of a snow-covered peak can fill our hearts with elation. And our eyes light up at the sight of bright red poppies stretching to the horizon, or brilliant sunshine flooding a length of sandy beach. We want to be part of places, feel their animate beauty touching our skin. Joy abounds in the waves that roll toward shore with operatic grace. The playful pull of the tide, the smell of faraway and underwater kingdoms, and the rumbling of the ocean remain in our souls long after the last grain of sand has washed away from our feet. Joy spills forth in late afternoon's shimmering light that bounces off the current of a rushing stream. How lovely to float in a lake so crystal clear it is possible to swim and drink at the same time. The blazing colors of the setting sun penetrate the pure, running creek. As the water bubbles over the rocky bed, a soothing bliss spreads over the earth with dusk's approaching silence, broken only when the small stones roll and dance, creating a lively melody of blurred motion. An invigorating hike is rewarded by trembling wildflowers that light up a meadow in vivid violet and yellow hues as far as the eye can see. To bask under the warm summer sun, lost in the beauty and drama of nature—that is joy.

AJIT VERMA
Perfect symmetry and design create a thrill of delight in this close-up of a bright orange flower.

CRISTINA MITTERMEIER
A late afternoon bath turns into a joyful water fight in the waters of the Iriri River, Brazil.

following pages
NIGEL HICKS
Spanning the landscape, a rainbow arches across marshland on Chiloé Island, Patagonia, Chile.

Brightly colored poppies dance in the strong
breeze of Big Sur, California.

PETE OXFORD
A Verreaux's sifaka hops from tree to tree across open ground in the spiny desert of southern Madagascar.

Find out where joy resides,
and give it a voice far beyond singing.
For to miss the joy
is to miss all.

—ROBERT LOUIS STEVENSON

The sun does not shine
for a few trees
and flowers, but for
the wide world's joy.

—HENRY WARD BEECHER

JOHN BURCHAM
As if suspended in midair, a tumbleweed is swept by the wind over the Bonneville Salt Flats, Utah.

preceding pages
FRANS LANTING
Thousands of pure white snow geese land in Klamath Basin National Wildlife Refuge, California, after a long flight.

One touch of nature

makes the

whole world kin.

—WILLIAM SHAKESPEARE

VICTOR MAZZARA

A dandelion's delicate seeds fan out like cats' whiskers.

Hope is the thing with feathers
That perches in the soul,
And sings the tune without words,
And never stops at all.

—EMILY DICKINSON

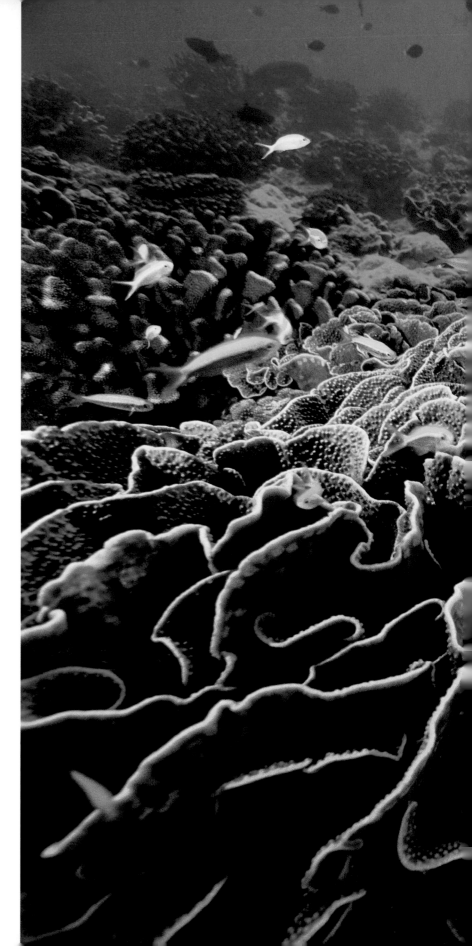

PAUL NICKLEN
A school of fairy basslets darts to and fro among lettuce coral blooms in the Phoenix Islands in the central Pacific.

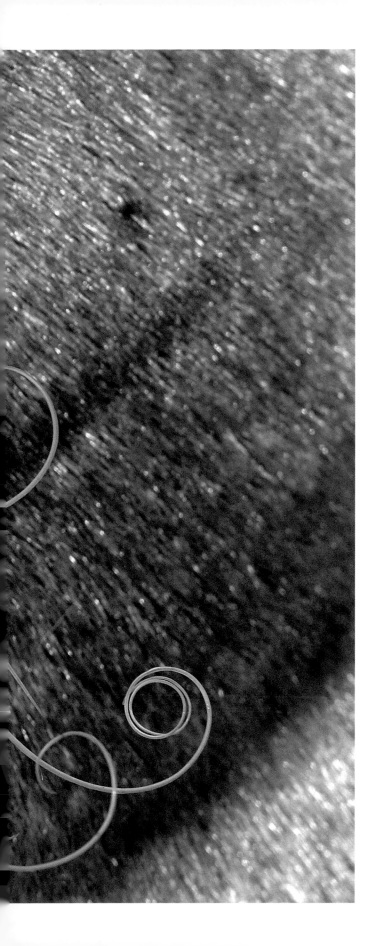

To see a world in a grain of sand,

And a heaven in a wild flower,

Hold infinity in the palm of your hand,

And eternity in an hour.

—WILLIAM BLAKE

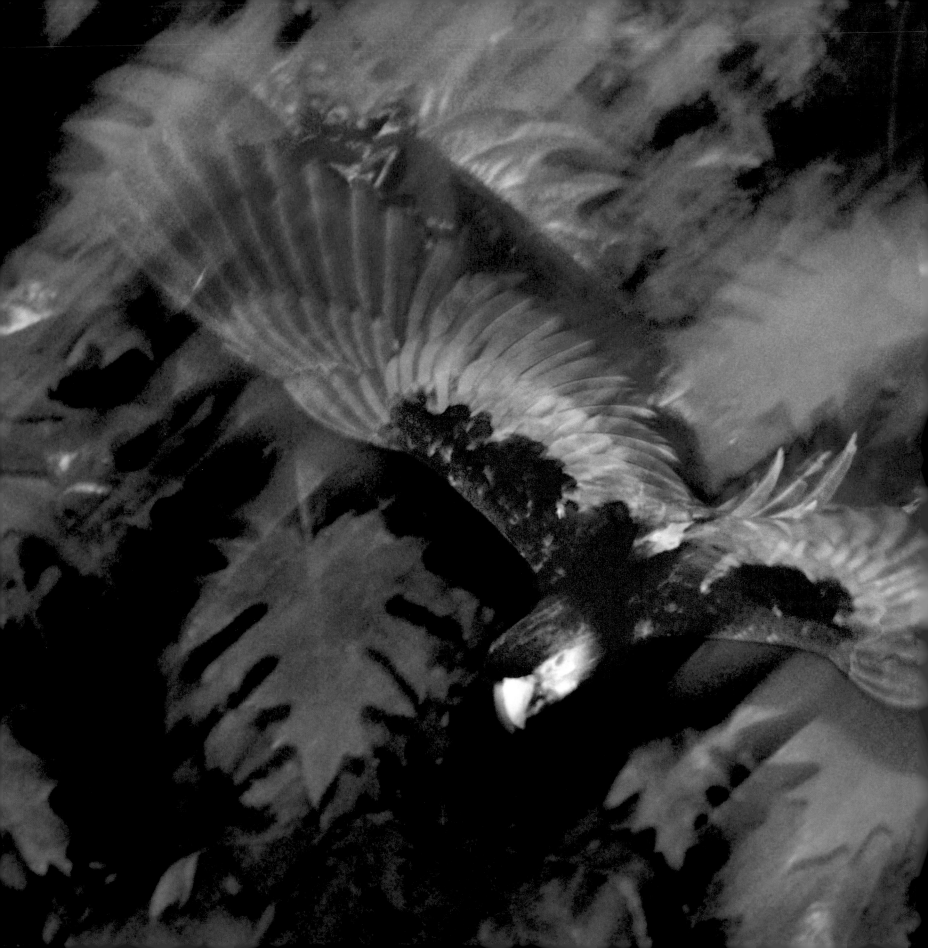

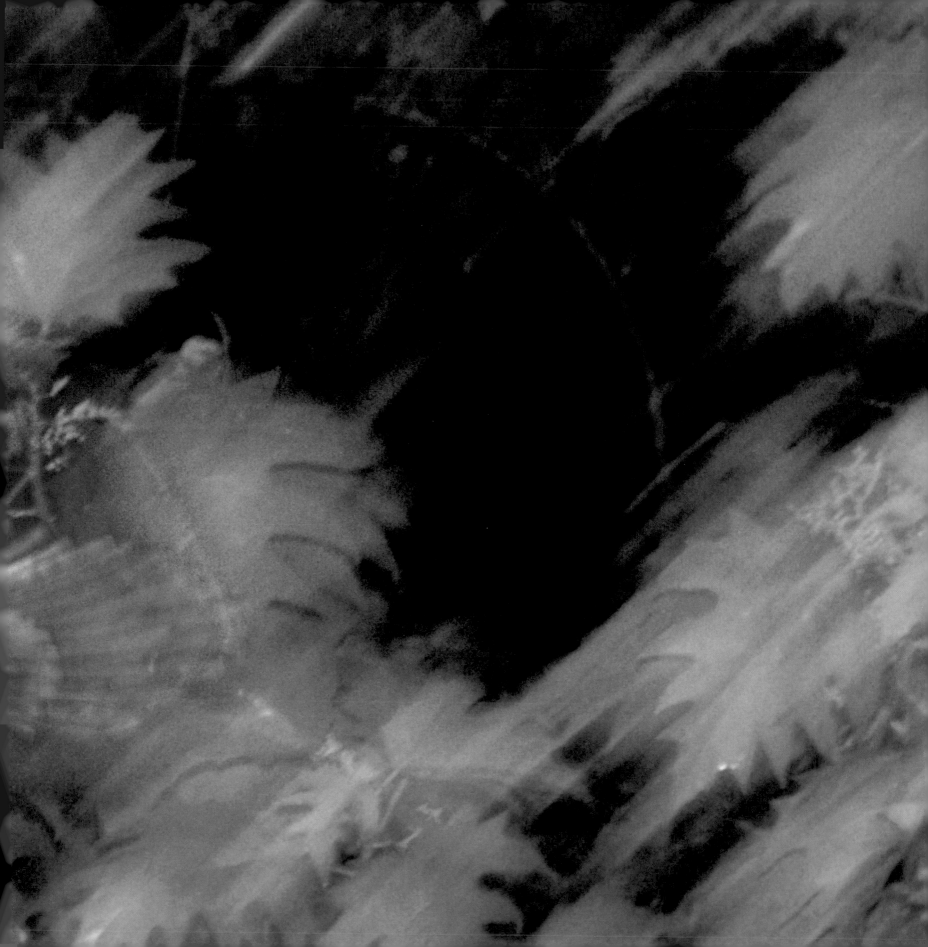

TOM TILL
A bright shaft of light penetrates the early morning mist in the Grand Canyon, Arizona.

preceding pages
FRANS LANTING
In a glorious flurry of red and blue feathers, a scarlet macaw takes flight in Tambopata National Reserve, Peru.

PAUL NICKLEN
An adult leopard seal uses a large chunk of ice to scan its surroundings in the Antarctic Peninsula.

DAVID DOUBILET

A goby hides among branches and polyps of gorgonian coral off the coast of Papua New Guinea.

With every drop of water you drink,
every breath you take,
you're connected to the sea.
No matter where on Earth you live.

—SYLVIA EARLE

FRANS LANTING
A steady wind sways the curious tufts of a grass tree in Western Australia.

following pages
PAUL NICKLEN
Huddling in tight groups to keep warm, adult king penguins pack together during a blizzard in St. Andrews Bay, South Georgia.

Shyly peeking out of a last spring snow, early blooming snowdrops light up the landscape in Germany.

Nature provides
exceptions to every rule.

—MARGARET FULLER

BILL CURTSINGER
Emerald waters provide a perfect backdrop for a green sea turtle in a Caribbean shoal.

following pages
WALTER MEAYERS EDWARDS
Morning sunshine backlights the thorns of cacti in the Organ Pipe Cactus National Monument, Arizona.

Lose yourself in Nature
and find peace.

—RALPH WALDO EMERSON

here the last road ends and the buzz of civilization dies down, that is where the solitude of nature begins. To reach that place we must journey, not only down the path that leads us away from all that is busy and worrisome, but also through the wilds of the human spirit. At our deepest core lies the sublime stillness of eternity. When we practice being in touch with our inner calm, it becomes habit and in tune with the friendly nature all around us. — Why not sit down beside a glassy blue lake and gaze up at the seamless blue sky and let nature flood you with peace? See how the gentle slope of a hill invites you to open your senses and revel in the myriad details around you. Wait for the late afternoon's cooling breeze and then amble down a grassy meadow. Every living thing stirs, sings, chirps, or calls; every blade of grass trembles as it soaks up the last rays of sun. Under nature's spell you can feel your spirit lifted and your soul healed. The collective harmony between soil, life, and weather reigns serene, and every humble living creature contributes to the scene, the fulfillment. In that grove of pines, a woodpecker will carve a nest; under its shade, scented by a thick blanket of needles, a trailing vine will grow, and on windy days, the branches of the pines will whisper a song. Our ears, our eyes, our very breath are part of it all—life and creation. — As the gathering dusk settles over the land, a great horned owl calls through the stillness of the dark—its longing tone reminds you of something you already know: Lucky moments spent in nature are rewarded with silence, calm reflection, and inner harmony.

CARR CLIFTON
The ancient trunk of a California live oak is covered with a thick crust of moss and lichen in Mount Diablo State Park, California.

CARR CLIFTON
A peaceful reflection of cattails creates a portrait of serenity at Mono Lake, California.

following pages
ART WOLFE
A lone gemsbok surveys its surroundings in Namib-Naukluft National Park, Namibia.

NAM PRASAD SATSANGI
A woman carrying water strolls through a beautiful mustard field in Punjab, India.

If you do not
expect the unexpected
you will not find it,
for it is not to be reached
by search or trail.

—HERACLITUS

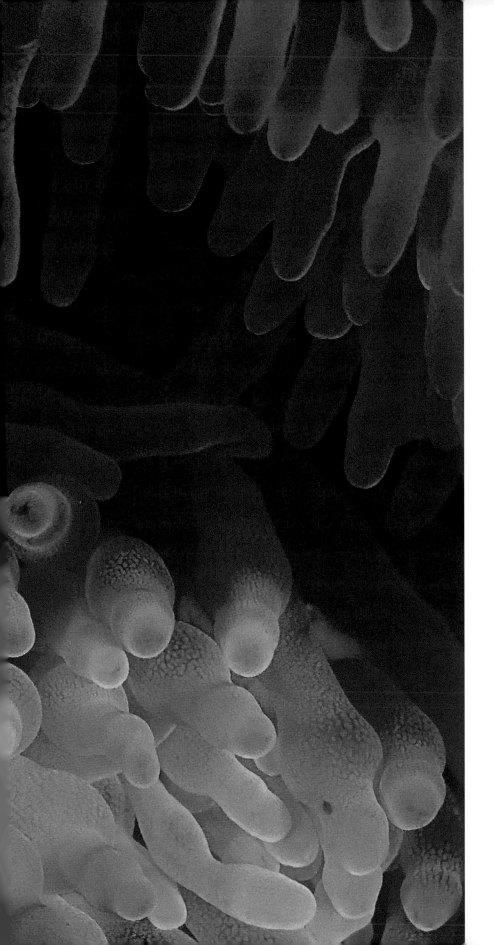

Going to the woods

is going home.

—JOHN MUIR

GALEN ROWELL
The first rays of sun set Owens Valley ablaze in California's eastern Sierra.

preceding pages
ART WOLFE
As if floating in air, piraputanga fish swim in clear waters in Brazil's Pantanal.

To secure a contented spirit,
measure your desires by your fortune,
and not your fortune by your desires.

—JEREMY TAYLOR

DAVID DOUBILET
*As if standing on the water's surface, a fisherman
casts his line into the clear waters of a coral islet
in the Raja Ampat Islands of Indonesia.*

preceding pages
MARIA STENZEL
*Late afternoon sun breaks through storm clouds
to illuminate Nevada's Las Vegas Valley.*

When we contemplate the whole globe

as one great dewdrop . . .

flying through space with other stars

all singing and shining together as one,

the whole universe appears

as an infinite storm of beauty.

—JOHN MUIR

RAUL TOUZON
As if performing an operatic ballet, a flock of pink flamingos strolls in a shallow lagoon in Celestún National Park, Yucatán, Mexico.

following pages
PAUL NICKLEN
Glassy designs emerge as pressure cracks sea ice in Canada's Arctic Circle.

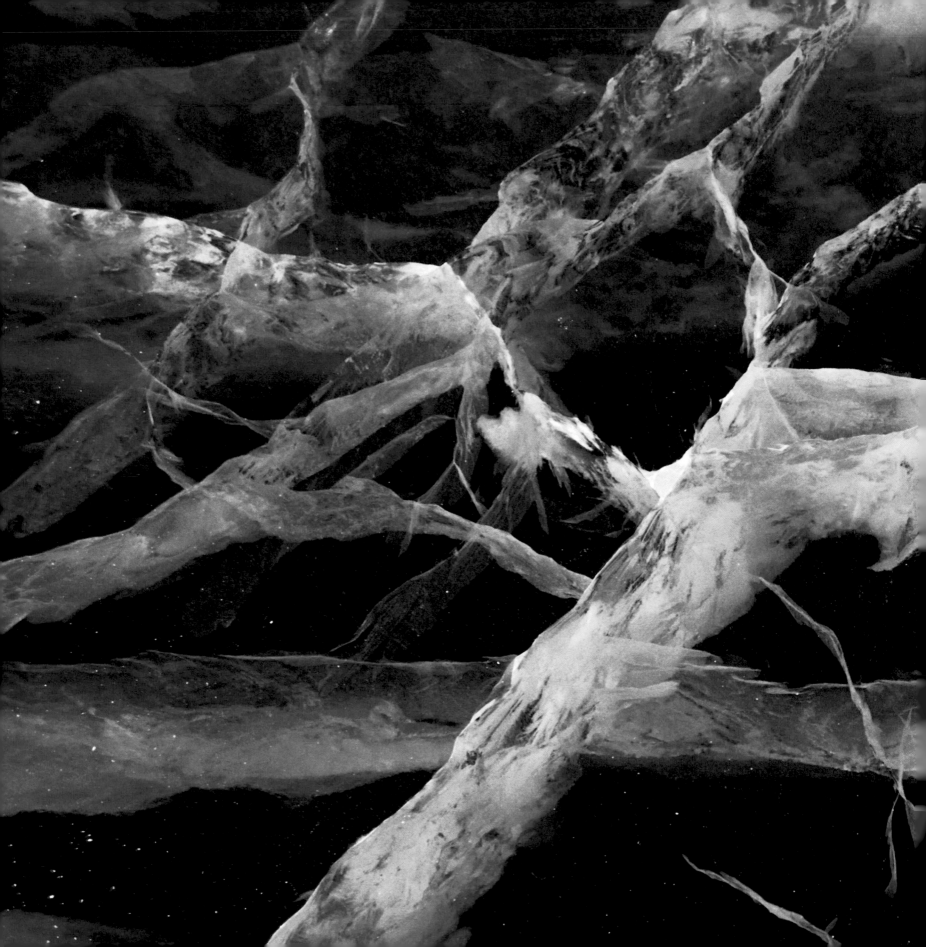

NORBERT ROSING
A napping polar bear in northern Canada is the picture of serenity.

A dreamer is one
who can only find his way by moonlight,
and his punishment
is that he sees the dawn
before the rest of the world.

—OSCAR WILDE

ROBBIE GEORGE
A sunset study of silhouettes and fiery skies appears as cormorants roost in a cypress tree at Lake Mattamuskeet, North Carolina.

GEORGE STEINMETZ
The ghostly figures of saguaro cacti stand under a starry sky in Bolivia's Uyuni salt flats.

MICHAEL MELFORD
A portrait of a delicate and ephemeral bog star flower in a meadow bordering a river in Kronotsky Zapovednik, Russia

The day the power of love
overrules the love of power,
the world will know peace.

—MAHATMA GANDHI

*Delicate, variegated sea urchin shells are for
sale in Portloe, Cornwall, England.*

Peace begins

with a smile.

—MOTHER TERESA

GARY MELNYSYN
Like a silent ghost, a wintering snowy owl takes flight in search of its next meal in Connecticut.

Peace comes
from within.
Do not
seek it without.

—GAUTAMA BUDDHA

Afterword

CAN BEAUTIFUL IMAGES INFLUENCE THE FATE OF OUR PLANET'S NATURAL WONDERS? The answer is a resounding yes. From the creation of the first national parks in the United States in the late 19th century to the more recent establishment of hundreds of protected areas around the world, photography has played a pivotal role in showcasing the beauty of wild places, the frailty of endangered species, and the wonder of indigenous cultures. Photography is, indeed, the pictorial voice of places and creatures that might otherwise be invisible. It has the power to mold perceptions, to stir our principles to action, and to inspire. Revealing such tremendous potential in an artistic medium has been an objective of this book. ⌒ As photographers, we have the choice to give our work a higher purpose, to invest it with values that reach beyond the physical beauty of a subject to awaken broad-based social consciousness. The imperative of safeguarding our planet's dwindling natural capital, and the urgency to change our current trajectory toward wildlife extinction, climate change, and civil conflict over diminishing resources, can be brought into sharper focus through photographic images. *Sublime Nature* is an invitation—a luscious one— to reflect on our individual connections to the natural world and also on the power of artistic expression to spur positive social change. In this way, the magnificent images in this book serve as ambassadors to a fast-changing and often careless world. In an enduring way, this volume brings hope for greater conservation and deeper appreciation for Earth's variety of landscapes. These pages and images send out an exuberant call for individual participation and also global involvement in keeping the gift of nature in and around us healthy and safe. There is an old proverb that says, "The best time to plant a tree is 20 years ago. The second best time is now."

DAVID DOUBILET
Mysterious and ghostly, two stingrays cruise the shallow, sandy sea bottom of the Cayman Islands' North Sound.

Acknowledgments

The production of a book never happens overnight and is never achieved in a vacuum. Always an invisible team keeps the wheels turning to make sure every *i* is dotted and every *t* is crossed. I would like to thank the amazing editors, designers, and staff in National Geographic's book division for their hard work, encouragement, and support. Janet Goldstein, senior vice president and editorial director of the book division, launched the project and offered keen advice and support throughout the production process. Senior editor Hilary Black lent her creativity and editorial acumen to shaping the book and ensuring that all parts came together in a work of perfection. No one should attempt to write anything in the absence of a great editor. I was lucky to work with the talented Gail Spilsbury, who gently massaged every word into beautiful text and a cohesive piece. As much as we love the creative process of making a book, there are endless tedious details that need tending to, and Bill O'Donnell was the quiet tinkerer who took care of all the necessary business pieces. A very special and heartfelt thanks to Sanaa Akkach, who put every ounce of her creative genius into this book, and to Meredith Wilcox, who lent her vast experience in photo editing to the image selection. My sincere gratitude to Marianne Koszorus for her artistic oversight of the project. The assistance of Myles Legacy with preparation of the manuscript was greatly appreciated. I have enjoyed the trust and encouragement of the CEMEX family for more than 20 years, and I would especially like to thank Luis Farías, senior vice president of energy and sustainability, and Vicente Saisó, corporate director for sustainability, for once again entrusting to me the production of their annual book. I also want to thank Alejandro Espinoza Treviño, conservation expert, for helping me to keep the project on track. I want to thank the photographers who have lent their talent and vision to this book. Their images are a visual language that crosses barriers, cultures, and languages. Their work connects people from all walks of life in every corner of the planet. Finally, I want to thank my family for being my source of inspiration—my children, John, Michael, and Juliana Mittermeier, and my partner, Paul Nicklen—for believing in me and helping me achieve every goal I ever set for myself.

Illustrations Credits

ASOCIACIÓN MUNDO AZUL, PERU: 110-111

AURORA PHOTOS: 186-187 (Mountain Light/Stock Connection)

CORBIS: 70, 90-91

GETTY IMAGES: 74-75

NATIONAL GEOGRAPHIC CREATIVE: 4-5, 6-7, 8-9, 10-11 (Minden Pictures), 12-13, 24-25, 28-29, 30-31, 34-35, 37, 38-39, 40-41 (Speleoresearch & Films), 42-43, 46-47, 48-49, 50-51, 52-53, 54-55, 56-57 (Minden Pictures), 58-59, 60-61, 62-63, 66-67, 76-77 (Minden Pictures), 84-85, 92-93, 94, 96-97, 100-101, 102-103 (Minden Pictures), 104-105, 106-107, 112-113, 124-125, 126-127, 128 (Minden Pictures), 134-135, 136-137, 138-139, 144-145, 146-147, 149, 150-151, 154-155, 157, 158-159, 160-161, 162-163 (Minden Pictures), 164-165, 166-167, 178-179, 182-183, 188, 190-191, 192-193, 195, 196-197, 198-199, 200, 202-203, 204-205, 206, 208-209, 210-211, 214-215 (Minden Pictures), 216-217, 218

NATIONAL GEOGRAPHIC YOUR SHOT: 2-3, 20, 27, 44, 114-115, 116-117, 120, 132-133, 140-141, 142, 176, 180-181, 212-213

Index of Photographers

A

SAM ABELL 208–209
MARC ADAMUS 22–3, 83
ALASKA STOCK 52–3
THEO ALLOFS 10–11
STEFAN AUSTERMÜHLE 110–11

B

TOM BEAN 109
JIM BRANDENBURG 102–103, 214–15
JOHN BURCHAM 136–7

C

IAN CAMERON 80–81
CARR CLIFTON 14–15, 86–7, 130–31, 170, 172–3
MIGUEL COSTA 114–15
BILL CURTSINGER 164–5

D

BILL DALTON 140–41
KENT DAVIS 27
DAVID DOUBILET 50–51, 157, 192–3, 218

E

JOHN EASTCOTT & YVA MOMATIUK 96–7
WALTER MEAYERS EDWARDS 166–7

F

NICK FORNARO 44

G

GORDON GAHAN 66–7
ROBBIE GEORGE 54–5, 202–203
JOSÉ LUIS GÓMEZ DE FRANCISCO 64–5

H

ROBERT B. HAAS 46–7
MAURICIO HANDLER 42–3
MARTIN HARVEY 90–91
NIGEL HICKS 124–5
MARK HOFER 2–3

J

MACIEK JABLONSKI 132–3

K

KOICHI KAMOSHIDA 74–5

L

TIM LAMAN 12–13, 100–101, 112–13, 138–9, 188
FRANS LANTING 4–5, 24–5, 104–105, 126–7, 134–5, 150–51, 158–9, 182–3

M

VICTOR MAZZARA 142
MICHAEL MELFORD 206
GARY MELNYSYN 212–13
CRISTINA MITTERMEIER 122–3
GEORGE F. MOBLEY 149
YVA MOMATIUK & JOHN EASTCOTT 96–7

N

CHRIS NEWBERT 76–7
PAUL NICKLEN 28–9, 30–31, 34–5, 38–9, 58–9, 84–5, 92–3, 144–5, 146–7, 154–5, 160–61, 198–9

O

PETE OXFORD 128

P

CARSTEN PETER 40–41, 48–9
JOE PETERSBURGER 94, 106–107
MICHAEL POLIZA 62–3

R

JIM RICHARDSON 210–11
NORBERT ROSING 200
GALEN ROWELL 186–7

S

JOEL SARTORE 6–7, 60–61
NAM PRASAD SATSANGI 176
JAMES L. STANFIELD 195
GEORGE STEINMETZ 37, 204–205
MARIA STENZEL 8–9, 190–91
HANS STRAND 70

T

TONY TAM 116–17
TOM TILL 152–3
MAREE TOOGOOD 20
RAUL TOUZON 178–9, 196–7, 216–17

V

AJIT VERMA 120

W

LAURA CRAWFORD WILLIAMS 98–9
BIRGITTE WILMS 56–7
ART WOLFE 32–3, 72–3, 78–9, 88–9, 174–5, 184–5
KONRAD WOTHE 162–3

Y

TUNC YAVUZDOGAN 180–81

Sublime Nature

Photographs That Awe and Inspire
Cristina Mittermeier

Published by the National Geographic Society

John M. Fahey, *Chairman of the Board
and Chief Executive Officer*

Declan Moore, *Executive Vice President; President,
Publishing and Travel*

Melina Gerosa Bellows, *Executive Vice President;
Chief Creative Officer, Books, Kids, and Family*

Prepared by the Book Division

Hector Sierra, *Senior Vice President and General Manager*

Janet Goldstein, *Senior Vice President and Editorial Director*

Jonathan Halling, *Design Director, Books and
Children's Publishing*

Marianne R. Koszorus, *Design Director, Books*

Hilary Black, *Senior Editor*

R. Gary Colbert, *Production Director*

Jennifer A. Thornton, *Director of Managing Editorial*

Susan S. Blair, *Director of Photography*

Meredith C. Wilcox, *Director, Administration
and Rights Clearance*

Staff for This Book

Gail Spilsbury, *Project Editor, Text Editor*

Sanaa Akkach, *Designer*

Marshall Kiker, *Associate Managing Editor*

Judith Klein, *Production Editor*

Mike Horenstein, *Production Manager*

Galen Young, *Rights Clearance Specialist*

Katie Olsen, *Production Design Assistant*

Production Services

Phillip L. Schlosser, *Senior Vice President*

Chris Brown, *Vice President, NG Book Manufacturing*

George Bounelis, *Vice President, Production Services*

Nicole Elliott, *Manager*

Rachel Faulise, *Manager*

Robert L. Barr, *Manager*

The National Geographic Society is one of the world's largest nonprofit scientific and educational organizations. Founded in 1888 to "increase and diffuse geographic knowledge," the member-supported Society works to inspire people to care about the planet. Through its online community, members can get closer to explorers and photographers, connect with other members around the world, and help make a difference. National Geographic reflects the world through its magazines, television programs, films, music and radio, books, DVDs, maps, exhibitions, live events, school publishing programs, interactive media, and merchandise. *National Geographic* magazine, the Society's official journal, published in English and 38 local-language editions, is read by more than 60 million people each month. The National Geographic Channel reaches 440 million households in 171 countries in 38 languages. National Geographic Digital Media receives more than 25 million visitors a month. National Geographic has funded more than 10,000 scientific research, conservation, and exploration projects and supports an education program promoting geography literacy. For more information, visit www.nationalgeographic.com.

For more information, please call 1-800-NGS LINE
(647-5463) or write to the following address:

National Geographic Society
1145 17th Street N.W.
Washington, D.C. 20036-4688 U.S.A.

For information about special discounts for bulk purchases, please contact
National Geographic Books Special Sales: ngspecsales@ngs.org

For rights or permissions inquiries, please contact
National Geographic Books Subsidiary Rights: ngbookrights@ngs.org

ISBN 978-1-4262-1301-4
ISBN 978-1-4262-1336-6 (special edition)
ISBN 978-1-4262-1327-4 (Spanish special edition)

Library of Congress Cataloging-in-Publication Data

Mittermeier, Cristina G.
[Photographs. Selections]
Sublime nature : photographs that awe and inspire / Cristina Mittermeier.
p. cm.
ISBN 978-1-4262-1301-4 (hardcover : alk. paper)
1. Nature photography. I. Title.
TR721.M58 2014
779'.3--dc23
2013020691

Printed in China
13/RRDS/1